The Euclid Space Telescope:

Revealing the Hidden Mysteries of Dark Matter and Dark Energy

Jeffery J. Westly

**Copyright © 2023 Jeffery J. Westly
All rights reserved.**

This work is protected by copyright law and may not be reproduced, distributed, transmitted, displayed, published, or broadcast without the prior written permission of the copyright holder.

Unauthorized use, duplication, or dissemination of this work is strictly prohibited and may result in legal action.

Table of Content

Introduction

Chapter 1: Exploring the Universe's Dark Side

Chapter 2: The Birth of the Euclid Space Telescope

Chapter 3: Euclid's Technological Marvels

Chapter 4: Observing the Cosmos: How Euclid Works

Chapter 5: Journey to the Dark Universe

Chapter 6: Data, Discoveries, and Scientific Breakthroughs

Chapter 7: Euclid and the Future of Astrophysics

Chapter 8: The Human Side of Space Exploration

Chapter 9: Beyond Euclid: Exploring the Universe's Frontiers

Conclusion

Introduction

Welcome to the World of Euclid

The universe is still guarding its deepest secrets in the obscurity of space and time, out there in the vastness of the cosmos. These puzzles include the enigmas of dark matter and dark energy, two enigmatic substances that together make up the vast bulk of our universe. Welcome to Euclid's intriguing universe, the space telescope on a mission to solve these cosmic mysteries.

The Search to Discover Dark Energy and Dark Matter

As we now understand it, the cosmos is a complex fabric of galaxies, stars, and cosmic structures that dance over the ages. However, there is much more to the cosmos than first seems. Astronomers and physicists made important discoveries in the 20th century that challenged our conception of the universe. Contrary to expectations, the cosmos is expanding faster than ever. The idea of dark energy, an unidentified force tearing the cosmos apart, was proposed by physicists to explain this occurrence. Galaxies and galaxy clusters also show gravitational tendencies that are not consistent with the observable matter alone. This inspired the idea of dark matter, a puzzling material that interacts gravitationally yet is hidden from view by our telescopes.

Since then, one of the biggest problems in the study of astrophysics and cosmology has been the effort to comprehend dark matter and dark energy. To learn more about the dark universe, the Euclid Space Telescope is a crucial tool in this project. In addition to being a scientific wonder, it represents human curiosity, inventiveness, and willpower.

A brief description of the Euclid Space Telescope Mission

The European Space Agency (ESA) has started work on the expansive Euclid Space Telescope. It is a component of ESA's Cosmic Vision program, which aims to explore the basic issues surrounding the cosmos. Euclid's main goal is to make accurate measurements of the universe's expansion and galaxy distribution in order to solve the riddles of dark matter and dark energy. The Visible Imaging Channel (VIS) and the Near Infrared Spectrometer and Photometer (NISP) are two of the state-of-the-art equipment it is furnished with. Euclid can map and quantify cosmic structures on a huge scale thanks to these techniques.

Euclid's approach is based on baryonic acoustic oscillations and gravitational lensing. The telescope may determine the distribution of stuff in the universe by monitoring the gravitational lensing's distortion of light from far-off galaxies. On the other hand, a special cosmic ruler provided by baryonic acoustic oscillations enables Euclid to determine the pace of the universe's expansion.

This Book's Structure and Goals

This book's objective is to introduce you to the world of Euclid, dark matter, and dark energy and to educate you along the way. The history of the Euclid Space Telescope, its technical advancements, and the complex science that underlies its functioning will all be covered. We shall experience the information, findings, and scientific advances that Euclid made in the realm of cosmology together. We will also explore the human aspect of space travel by getting to know the engineers and scientists who have committed their whole careers to this endeavor.

Also discussed in this book are the results of the Euclid Space Telescope and their ramifications. We will think about the theoretical implications of Euclid's findings and how they affect how we see the universe. The significance of Euclid's legacy extends beyond, affecting both the present and the future of astronomy and space travel as well as our understanding of the cosmos.

You will get an understanding of the difficulties, successes, and unyielding spirit of human inquiry as we go on this adventure together. This book is intended to be educational and approachable for all readers, whether they are seasoned astrophysicists, astronomy enthusiasts, or just naturally interested people. It presents complicated concepts in an interesting and intelligible way.

So buckle up and get ready to be led by the Euclid Space Telescope as you go far into the universe. Together, we shall solve the riddles of dark matter and dark energy, embracing the great grandeur of the cosmos in the process. In the next chapter, we dive into the fascinating universe of dark matter and dark energy, which ushers in the journey.

Chapter 1: Exploring the Universe's Dark Side

The universe has always captured the attention of people due to its vast number of galaxies and stars. But when we began our exploration of the cosmos and stared up at the night sky, we quickly discovered that there was more to the tale than what first seemed. Our investigation into the mysterious worlds of dark matter and dark energy, which are waiting for us to investigate, begins in this chapter.

The mystery of dark matter

The building blocks of galaxies and stars, or the matter, are the threads that make up our universe. However, when we study the motions of galaxies and galaxy clusters, a remarkable fact emerges: there is insufficient observable matter to explain the gravitational forces at work. We cannot see enough stars and galaxies to produce the gravitational effects seen in cosmic structures.

Due to this oddity, the existence of dark matter—an enigmatic element that is both invisible and mysterious and accounts for around 27% of the universe's total mass and energy—was postulated. Dark matter is invisible to the naked eye because it does not emit, absorb, or reflect light. Instead, the gravitational interactions it has with material in the observable universe provide evidence for its existence.

According to scientists, dark matter is made up of unusual particles that are different from the particles we normally come into contact with. They are known as Weakly Interacting Massive Particles (WIMPs), and it is believed that they wander across the cosmos, generating enormous cosmic halos surrounding galaxies. They exert gravitational pull on galaxies, changing the way they move and how they are structured. One of the most important unanswered questions in astrophysics is the make-up of dark matter.

Explicitly Defining the Dark Energy's Nature

Dark energy has a distinct but no less significant function to play in the overall story of the universe than dark matter, which is the unseen substance that affects galaxies on a cosmic scale. Despite assumptions that gravitational forces would operate to draw the universe together, scientists discovered in the late 20th century that it was expanding faster than it was contracting. Dark energy, a force that defies gravity and pulls galaxies apart, was believed to be responsible for this acceleration.

Even more enigmatic than dark matter, dark energy makes up around 68% of the universe's total mass and energy. It interacts via no known force, including gravity, unlike dark matter. The gravitational attraction of stuff in the universe is resisted by its behavior as a negative pressure. One of the most difficult problems in modern physics and cosmology is to understand the nature of dark energy, which remains a deep mystery.

Why Dark Matter and Dark Energy Are Important

It is a mission to unravel the fundamental fabric of our cosmos; understanding dark matter and dark energy is not just an academic endeavor. The solution to understanding the past, present, and future of the cosmos lies in these mysterious elements.

The large-scale architecture of the universe are caused by dark matter and its gravitational pull. It gives galaxies a cosmic framework on which to stand and allows them to revolve without splintering. Our universe would be a very different place without dark matter, and the galaxies that we know and love today would not exist.

The future of the cosmos is instead shaped by dark energy. The cosmos is expanding faster due to its repellent force, and the kind of force it emits determines whether the universe is flat, open, or closed. The characteristics of dark energy are connected to the destiny of the universe, from a Big Crunch through endless expansion.

Historical Viewpoint

The search for answers to the mysteries of dark matter and dark energy did not start off suddenly. It is the result of decades' worth of theoretical development, technical advancements, and astronomical observations. Our current knowledge of the universe is the result of the ground-breaking work of scientists like Albert Einstein, Georges Lemaître, and Edwin Hubble.

Albert Einstein made the disastrous choice to establish the idea of the cosmological constant, which he subsequently described as his "biggest blunder." When Hubble's discoveries revealed an expanding cosmos, the constant that had been previously postulated to sustain a static universe was abandoned. In recent years, nevertheless, it has come back into favor as a possible justification for dark energy.

The concept of an expanding cosmos and what would come to be known as the Big Bang hypothesis were presented by Belgian physicist and Catholic priest Georges Lemaître. His contributions laid the foundation for our knowledge of cosmic expansion and the universe's creation.

The cosmos is expanding, as shown by Edwin Hubble's studies of far-off galaxies and their redshifts. The cornerstone for contemporary cosmology was set by this revelation, which altered our cosmic viewpoint.

Chapter 2: The Birth of the Euclid Space Telescope

We look at the intriguing history of how the Euclid Space Telescope came to be as we set out on our quest to investigate the mysteries of dark matter and dark energy. The Euclid Mission's origins, main goals, place in the Cosmic Vision Program of the European Space Agency (ESA), as well as the cooperative efforts and international collaborations that made it possible, are all covered in this chapter.

Beginning of the Euclid Mission

A critical desire to explore further into the cryptic worlds of dark matter and dark energy gave rise to the Euclid Mission, which was developed at the beginning of the twenty-first century. The European Space

Agency, a major contributor to astronomical study and space travel, is where the project was conceived.

The ESA planned to create a specialized mission to unravel the secrets of the universe's dark side since it was urgently necessary to comprehend these cosmic riddles. The Euclid Mission would come to exemplify this ideal, bringing together cutting-edge technology, global collaboration, and scientific inquiry.

Important aims and objectives

There are several important aims and goals for the Euclid Mission, each designed to increase our understanding of dark matter and dark energy. These goals consist of:

One is the mapping of dark matter. The main goal of Euclid is to map the location of dark matter with the highest level of accuracy possible. The telescope seeks to construct a thorough picture of the cosmic web generated by dark matter by examining how gravity bends and distorts light from distant galaxies.

2. Examining Dark Energy In addition to deciphering the mysteries of dark matter, Euclid's task is to look into dark energy. The telescope will shed light on the enigmatic force that opposes gravity by viewing the accelerating expansion of the cosmos.

3. Astronomical Surveys Galaxies, galaxy clusters, and their cosmic distribution will all be thoroughly surveyed by Euclid. These surveys will make it easier to comprehend the universe's large-scale structure and development.

(4) Multiband Imaging Precision observations in several wavelength bands are made possible by the telescope's visible

and near-infrared imaging sensors. The characteristics of dark matter and dark energy must be understood using this multiband methodology.

5. Cosmological History: Euclid's observations will go back in time and enable astronomers to investigate cosmic history, from the beginning of the universe to the present.

The Cosmic Vision Program of the ESA and Euclid

The Euclid Mission is in line with the Cosmic Vision Program of the ESA, a long-term strategy that outlines the organization's objectives in space research and exploration. Aiming to answer basic questions about the cosmos, its beginnings, and its future, the Cosmic Vision Program sets ambitious goals. As a medium-class mission that supports the program's broader goals, Euclid has an important role within it.

The Cosmic Vision Program was established in 2015 and focuses on four scientific areas:

1. "Hot and Energetic Universe": Examining very high-energy phenomena including neutron stars, black holes, and the universe's extreme temperatures.

2. "The Gravitational Universe": looking into the origins of gravity and gravitational waves, as shown by initiatives like LISA (Laser Interferometer Space Antenna).

3. The Cosmic Microwave Background: Investigating the Big Bang's afterglow to learn about the circumstances and history of the early cosmos.

4. Planetary Defense and Near-Earth Objects: Creating plans to lessen possible effects of asteroids on Earth.

The second focus of the program is "The Gravitational Universe," and Euclid contributes considerably to our knowledge of both the universe's expansion and the role of dark matter and dark energy in determining its future.

International Partnerships and Collaborative Efforts

The success of the Euclid Mission is proof of the value of teamwork and global cooperation in the field of space research. A group of international partners and research institutes support the project, which is led by the ESA.

Principal participants are:

- The Euclid Consortium: The Euclid Consortium, which has over 1,000 members from 15 different European nations, brings together scientists, engineers, and researchers committed to the mission's success.

- NASA: NASA contributes the Near Infrared Spectrometer and Photometer (NISP) instrument to Euclid, which is largely an ESA project.

Industrial Partners: The spacecraft and equipment for Euclid were developed and built by many firms from ESA member states.

By expanding our knowledge of dark matter, dark energy, and the development of the cosmos, these collaborations show the worldwide effort put into the endeavor.

Chapter 3: Euclid's Technological Marvels

It's crucial to know the amazing technical miracles that drive the Euclid Space Telescope as we work to unravel the mysteries of dark matter and dark energy. This chapter explores the state-of-the-art equipment, spacecraft, and instruments that make Euclid's mission feasible.

Instruments and Hardware with Cutting Edge

Equipped with a variety of state-of-the-art tools and technology that enable very accurate observations of the cosmos, Euclid is a tribute to human creativity. The mission's main goals of mapping dark matter and examining dark energy have been methodically served by these technical wonders.

The Visible Imaging Channel (VIS) and the Near Infrared Spectrometer and Photometer (NISP) are two essential tools at the core of Euclid's scientific capabilities. Astronomers may investigate the universe in many wavelength bands thanks to the combination of these devices, which together record pictures and data.

(VIS) The Visible Imaging Channel

To get detailed pictures of galaxies and galaxy clusters, the Visible Imaging Channel (VIS) is a unique tool. To map the distribution of dark matter across the cosmos, VIS, which operates in the visible light spectrum, is very important. It does this by studying the gravitational lensing phenomenon, in which the gravitational pull of large objects like galaxy clusters bends and amplifies light from background galaxies.

A large visible-wavelength camera with a broad-band filter and a spectroscopic channel with capabilities for low- and medium-resolution spectroscopy make up the VIS system. This adaptable tool enables astronomers to carry out a variety of investigations, from precise galaxy photography to the spectrum analysis of their light.

For Euclid's main objective, the pictures acquired by VIS are crucial because they allow for the development of exact maps that show the cosmic web of dark matter and provide information about the large-scale features of the cosmos.

NISP, or the Near Infrared Spectrometer and Photometer

Another ground-breaking tool on board Euclid that works in conjunction with VIS is the Near Infrared Spectrometer and Photometer (NISP). The electromagnetic spectrum's near-infrared region is where NISP works, enabling it to collect spectroscopic data and pictures that are not visible to the human eye.

NISP is made to look at the rapid expansion of the cosmos to investigate dark energy. NISP can measure the impact of dark energy on the pace of universe expansion by analyzing the near-infrared light from distant galaxies. This device has a spectroscopic channel for detecting the redshift of galaxies as well as a photometric channel for imaging.

Together, VIS and NISP provide a comprehensive perspective of the universe that spans a variety of wavelengths and enables astronomers to address key issues including dark matter, dark energy, and cosmic architecture.

Vehicles and Operations in Space

The spacecraft itself is a tremendous technical achievement, in addition to the technological wonders found in Euclid's instruments. The spacecraft's architecture is focused on precise observations, reducing the effect of outside influences that can compromise the accuracy of the data. A sun shield on it keeps stray light out of the way and safeguards observations.

Euclid is located in the second Lagrange point (L2), a region in space that is advantageous for astronomical observations. This fixed location in space enables the telescope to track the Earth as it revolves around the Sun while keeping a constant distance from our planet. Euclid can monitor the same areas of the sky for prolonged periods as a result, providing complete survey field coverage.

To fine-tune its orientation and provide precise control over its observations, the telescope is outfitted with reaction wheels and thrusters. When Euclid collects data, it transmits it to Earth, where a team of scientists and researchers carefully examine and analyze it.

International partners, space agencies, and scientific institutions all work together to carry out Euclid's activities. Its success depends on the cooperation and knowledge of many people and groups devoted to understanding the secrets of the cosmos.

Chapter 4: Observing the Cosmos: How Euclid Works

The Euclid Space Telescope, a wonder of technological advancement, was created to shed light on the mysteries of dark matter and dark energy. This chapter will examine Euclid's complicated operations, including how it handled enormous quantities of astronomical data and mapped the cosmos.

The mapping of the cosmos

The challenging objective of Euclid's mission is to map the distribution of dark matter in the cosmos with the highest level of accuracy. Euclid does this by using the amazing gravitational lensing method. Einstein's general theory of relativity predicts that enormous objects, like galaxy clusters, may bend and distort the course of light, which leads to the phenomena known as gravitational lensing. This lensing effect enlarges and warps the light from background galaxies, enabling astronomers to see and quantify the distribution of dark matter.

To map the cosmos via gravitational lensing, Euclid's Visible Imaging Channel (VIS) is essential. Astronomers can examine the minor distortions brought on by gravitational lensing thanks to the high-resolution photos it takes of galaxies and galaxy clusters. The large-scale features of the universe may be revealed by examining these aberrations, which allow researchers to produce exact maps of the distribution of dark matter.

Euclid makes use of the idea of baryonic acoustic oscillations (BAO) in addition to gravitational lensing. BAO refers to recurring changes in the density of baryonic matter (common stuff, such as protons and neutrons), which were left behind in the early cosmos. Astronomers can gauge the history of the universe's expansion using these oscillations as a cosmic ruler. The cosmic web is fully understood by Euclid by combining gravitational lensing with BAO data.

Graceful Lensing and Baryonic Acoustic Oscillations

Euclid uses two potent methods to probe the dark side of the universe: gravitational lensing and BAO. While BAO data aid in the understanding of the universe's expansion rate, gravitational lensing allows the telescope to study dark matter structures. Together, these methods provide a comprehensive understanding of cosmic evolution.

Astronomers may learn more about the characteristics and distribution of dark energy thanks to gravitational lensing, which also makes dark matter more visible.

Understanding the nature of dark energy is crucial since it is what propels the universe's accelerating expansion. By probing the enigmatic characteristics of dark energy and revealing its mysteries, Euclid can leverage the small distortions in galaxy forms brought on by gravitational lensing.

The history and pace of the universe's expansion may be learned through BAO observations. Euclid can accurately calculate the pace of cosmic expansion by examining the distribution of galaxies at various times in the universe's history. The dynamics of dark energy and its impact on the course of the universe are being studied by scientists with the aid of this data.

Observing the Sky

The purpose of Euclid's mission is to conduct a thorough survey of the sky that includes the whole cosmos. From the very beginning of the cosmos to the present, the telescope plans to study galaxies and galaxy clusters. This thorough study will provide researchers with a plethora of information to examine and make important discoveries regarding cosmic structures, dark matter, and dark energy.

To examine the sky, Euclid takes the following sequential steps:

1. Survey Fields: The telescope picks out certain swaths of the sky that are known as "survey fields." To optimize the mission's scientific effect, these fields have been carefully selected.

2. Observations: Euclid is capable of taking high-resolution photos and spectral data in the visible and near-infrared spectrums. It continually scans the survey fields, gathering a comprehensive dataset.

3. Data Collection: Euclid sends the information it has gathered to Earth, where

it is carefully processed and analyzed. As they carefully examine the data, researchers look for gravitational lensing effects and BAO signs.

4. Cosmic Cartography: Researchers utilize the data to produce exact maps of the distribution of dark matter and the characteristics of dark energy. These maps give information on the cosmos's development by revealing the large-scale architecture of the universe.

Data Handling and Processing for Astronomy

To handle and analyze the enormous quantities of astronomical data that Euclid was able to view and record, a complex system is required. The spacecraft carefully transfers the data it has gathered to Earth-based facilities, where a committed group of scientists and researchers works to handle the difficult process of managing the data.

Analysis, calibration, and data reduction are steps in the process. Astronomers carefully examine the pictures and spectra and make any necessary adjustments for instrumentation and ambient factors. For significant scientific research to be possible, calibration is essential to ensure the data's correctness.

To sum up, the observational techniques used by the Euclid Space Telescope, such as gravitational lensing and baryonic acoustic oscillations, allow it to map the cosmos while illuminating the mysterious nature of dark matter and dark energy. A strong instrument for solving the secrets of the cosmos is made possible by a thorough scan of the sky and meticulous data collection and processing.

Chapter 5: Journey to the Dark Universe

In this chapter, we set off on a remarkable trip with the Euclid Space Telescope as it launches, deploys, and travels across space while following its special orbital route to solve the riddles of dark matter and dark energy. This chapter gives readers a behind-the-scenes look at the mission's phases and the heroic efforts made to overcome technical obstacles.

Deployment and Launch

Launch and deployment are the first important steps in the Euclid mission's trek to the dark cosmos. Euclid was successfully launched into orbit on a Soyuz rocket on June 20, 2022, from the Guiana Space Center in French Guiana. The telescope's amazing journey was made possible by the launch, which was a critical turning point.

After a brief but arduous ascent into the sky, Euclid arrived at its intended orbit and was launched into space. For the telescope to start its mission of cosmic exploration, this step entailed the accurate release of the instrument onto its intended trajectory.

The Earth's trajectory and Euclid's orbital path

Euclid's orbital path was meticulously planned to maximize its astronomical observations. Euclid takes a special orbit known as an "Earth trail" compared to many other observatories that circle the Earth. Euclid will follow Earth as it circles the Sun thanks to this trajectory, which positions it in a precise position concerning Earth.

The mission benefits from various factors from the Earth-trail orbit. It reduces atmospheric and solar radiation interference, ensuring that Euclid's observations are unaltered and clear. This orbit also makes it possible for the telescope to see a large portion of the sky, aiding its thorough examination of galaxies and galaxy clusters.

Mission Phases (Stages)

There are many phases to the Euclid expedition, each with its own goals and difficulties. These phases comprise:

1. Commissioning Phase: Euclid underwent a commissioning phase after its launch and deployment. The telescope's equipment and instruments underwent extensive testing and calibration during this time to make sure they were operating as intended. This crucial stage is intended to get Euclid ready for its scientific research.

2. Scientific Operations: After the commissioning stage was successfully finished, Euclid started its main scientific

operations. The telescope started its systematic scan of the cosmos, taking pictures and collecting spectral information with a special emphasis on gravitational lensing and baryonic acoustic oscillations.

3. Data transmission and analysis: Euclid transmits its acquired data to Earth-based facilities for evaluation. Astronomers and scientists labor hard in teams to handle and analyze massive information. To get insightful knowledge about dark matter, dark energy, and cosmic structures, the analysis step is very important.

4. Continuous Observations: Euclid's mission is continuing, and it is always gathering data. A thorough scan of the universe is possible thanks to the telescope's protracted operation. For detecting subtle cosmic occurrences, a longer observation period is necessary.

Getting Past Technical Difficulties

There have been several technical difficulties with the Euclid expedition. A unique set of challenges must be overcome to go across space and collect accurate astronomical data.

Maintaining the stability of Euclid's sensors and detectors is one of the biggest technological issues it encounters. The performance of the telescope may be affected by radiation and temperature changes in the vacuum of space. Euclid's instruments are outfitted with cutting-edge cooling systems and radiation shielding to lessen these impacts and ensure the accuracy of its observations.

To take exact pictures and spectra, the spacecraft must also constantly change its attitude. Advanced control systems and propulsion systems work together to fine-tune Euclid's position and orientation, enabling the telescope to follow the trajectory it wants to.

A monument to our intellect and our unquenchable curiosity about the world is Euclid's trek into the black universe. The success of the project depends on cooperation between scientists, engineers, and space agencies who are all committed to solving the mysteries of dark matter and dark energy.

Chapter 6: Data, Discoveries, and Scientific Breakthroughs

The Euclid Space Telescope is gathering a plethora of data as it travels through space, which has the potential to shed light on some of the biggest mysteries in the cosmos. This chapter examines the first observations and findings of the Euclid project, with a particular emphasis on the limitations placed on dark energy, the search for dark matter structures, and the mission's overall effects on cosmology.

Early Observations and Findings

Since its launching and deployment, Euclid has carefully surveyed the cosmos, collecting spectral data and photos from a wide variety of galaxies and galaxy clusters. Early observations by Euclid have already yielded important new knowledge about the structure of the cosmos. The huge dataset that the telescope is anticipated to amass throughout its prolonged mission is only partially visible in these images.

The capacity to map and quantify the distribution of galaxies across the cosmos is one of Euclid's early discoveries' noteworthy features. Astronomers can better comprehend the universe's development and composition by looking at the cosmos' large-scale structure. Initial findings from Euclid's observations have revealed important details regarding the distribution of galaxies and the cosmic web, an intricate web of filaments and voids that constitutes the structure of the cosmos.

Restrictions on Dark Energy

The main goal of Euclid's mission is to learn more about dark energy, the unidentified factor that is speeding up the universe's expansion. Euclid's findings have put limitations on dark energy's characteristics, revealing insight into its actions and their cosmic ramifications.

The investigation of baryonic acoustic oscillations (BAO) is one of Euclid's major methods for examining dark energy. It is possible to gauge the pace of the universe's expansion using these tiny patterns in the distribution of galaxies. To determine if the effects of dark energy are compatible with the cosmological constant suggested by Albert Einstein or whether there are more complicated dynamics at work, scientists may determine the limitations on the equation of state of dark energy by carefully measuring the BAO scale.

According to preliminary findings, dark energy continues to fit the description of a cosmological constant put out by Einstein. However, Euclid's high-precision observations will give us a more complete knowledge of the nature of dark energy, and future data processing could make unexpected findings.

Dark matter structures: the search

Euclid is mainly concerned with dark energy, but it also has a significant impact on the continuing investigation into the presence of dark matter structures in the cosmos. It is impossible to detect dark matter using conventional methods because it does not interact with light or other electromagnetic forces. It can only be deduced from the gravitational effects it has on visible stuff, including galaxy clusters and galaxies.

By depicting the distribution of matter in the cosmos, Euclid's observations aid in this effort. Astronomers can map the distribution of dark matter because of gravitational lensing, a process where the gravitational pull of big objects bends the direction of light. Scientists may locate and comprehend the features of dark matter formations by analyzing the gravitational lensing effects in Euclid's data.

To comprehend the large-scale cosmic structure and the function of dark matter in the development of the universe, it is imperative to identify and characterize dark matter formations. Our understanding of the intangible cosmic framework created by dark matter will be improved by Euclid's contribution to this project.

Euclid's Influence on Cosmology

The purpose of Euclid is far broader than only studying dark matter and energy. The study of cosmology has been significantly impacted by its thorough examination of the cosmos. The observations of the telescope address a broad variety of cosmological issues, including the nature of dark energy and dark matter, the pace of universe expansion, and the development of cosmic structures.

Euclid's ability to provide accurate measurements for a range of cosmic parameters is one of his most important contributions to the field of cosmology. These observations aid in the improvement and verification of our cosmological models. The history, present, and future of the cosmos may be better understood by cosmologists by merging data from Euclid with other astronomical discoveries.

Euclid's influence on cosmology also extends to how he handled important issues like the universe's future. Euclid's data can reveal whether or not the universe will undergo a shift in its cosmic destiny by monitoring the expansion rate and the equation of the state of dark energy.

Early observations and findings from the Euclid Space Telescope are already altering our perception of the cosmos. It will continue to be a pillar of astronomical study for many years to come thanks to Euclid's contributions to the study of dark energy, dark matter structures, and cosmology in general.

Chapter 7: Euclid and the Future of Astrophysics

In particular, the areas of dark matter, dark energy, and large-scale cosmic structure have seen substantial advancements thanks to the Euclid Space Telescope. The final chapter of this book delves into Euclid's enduring legacy, the fascinating possibilities offered by upcoming space telescopes and missions, Euclid's contribution to the advancement of space exploration, and the enormous potential for new astrophysical discoveries that lie ahead.

The Influence of Euclid

Astrophysics will leave a significant and lasting legacy thanks to Euclid. It is anticipated that the understanding of dark energy, dark matter, and the large-scale cosmic structure of the universe will grow as the mission continues and more data is gathered. Future astronomical studies will be measured against the exact measurements and high-quality data that Euclid gave.

Contributing to our knowledge of dark energy is one of Euclid's lasting legacies. The limitations imposed on the dark energy equation of state will continue to affect cosmological models and the advancement of our understanding of the universe's accelerating expansion.

Furthermore, Euclid's contribution to the mapping of dark matter structures will deepen our knowledge of the unseen framework of the cosmos. Future studies and discoveries will benefit from this legacy's influence on the development and distribution of dark matter structures in the universe.

Missions and Space Telescopes to be Launched Soon

While Euclid is a significant advancement in our investigation of the cosmos, it is not the last frontier. There are several prospective space telescopes and missions in the subject of astrophysics that aim to deepen our knowledge of the universe. These missions will expand on Euclid's basis and explore more basic astrophysical topics.

One of the noteworthy planned missions is the James Webb Space Telescope (JWST), which is sometimes referred to be the Hubble Space Telescope's successor. The JWST, which is set to launch, is intended to monitor the cosmos in the infrared spectrum, enabling the study of hitherto unknown cosmic events. It is anticipated that it will look into the creation of stars and galaxies, the development of planetary systems, and the discovery of habitable extrasolar planets.

The Nancy Grace Roman Space Telescope (formerly known as WFIRST), which will scan the sky in the optical and near-infrared wavelengths, is another project that will soon launch. Expanding on the foundation Euclid built, Roman will continue to explore the mysteries of dark matter, dark energy, and cosmic acceleration.

The Euclid expedition will collaborate with these next projects, advancing our knowledge of the cosmos as a whole. A new age of astronomical discoveries will begin thanks to the cooperation between these telescopes, giving us a more complete picture of the universe.

The Contribution of Euclid to Space Exploration

In addition to expanding our understanding of the cosmos, Euclid was crucial in developing space exploration technology and capabilities. The state-of-the-art equipment, sensors, and spacecraft parts created for Euclid have uses that go beyond this mission and are advantageous to the next space initiatives.

The Near Infrared Spectrometer and Photometer (NISP) and Visible Imaging Channel (VIS), two of Euclid's technical wonders, have pushed the limits of observational astronomy. Future telescopes and space missions will be able to see the universe in greater depth and breadth thanks to the technologies that will serve as their foundation.

Aspects of Euclid's functioning, such as its orbital route and Earth-trail trajectory, can provide important insights into the organization and execution of missions. Future missions' plans and techniques will be influenced by the lessons learned during Euclid's trip through space, resulting in success and increased scientific output.

Opportunities for Potential Future Discoveries

The scientific community is looking forward to the chances for discoveries that lie ahead as Euclid continues its mission. The enormous amount of information Euclid gathered is a goldmine of possible discoveries and insights.

The hunt for exoplanets is one of the subjects ready for investigation. It is now possible to find and describe exoplanets, including those in habitable zones where life as we know it may exist, thanks to Euclid's exact measurements and thorough study of the universe. Such findings could alter how we think about the likelihood of life existing across the cosmos.

Furthermore, Euclid's contributions to cosmology will continue to provide light on the universe's future. Dark energy and dark matter research will probably find unanticipated events and characteristics that contradict current hypotheses and open up new lines of investigation.

The purpose of Euclid is a monument to human curiosity and our unrelenting search for knowledge about the universe. As a result of its history, the potential of prospective space telescopes and missions, its contribution to furthering space exploration, and the many prospects for discoveries, Euclid will continue to be a leader in astrophysics study for many years to come. Not only have the secrets of dark matter and dark energy been revealed by the Euclid Space Telescope, but it has also paved the way for further investigations into the mysterious depths of the cosmos.

Chapter 8: The Human Side of Space Exploration

The Euclid Space Telescope is a technological wonder and a symbol of advancement in science, but its instruments and data are only possible thanks to the tireless work of many scientists and engineers. This chapter delves into the human side of space exploration, honoring the people who made Euclid a reality, the spirit of global collaboration that made this mission possible, and the individual tales, anecdotes, struggles, and victories that have shaped Euclid's journey.

Who Developed Euclid: The Engineers and Scientists

Without the great brains who came up with the mission's concept, design, and construction, Euclid's voyage into the depths of the cosmos would not have been feasible. To bring the idea of Euclid to reality, a diverse group of scientists, engineers, and technicians from across the world gathered.

These hard working professionals contributed their knowledge in a variety of disciplines, including optics, spacecraft engineering, cosmology, and astrophysics. From creating cutting-edge equipment capable of exact observations to assuring the spacecraft's flawless launch and deployment, they faced enormous obstacles. The strength of human curiosity and inventiveness is shown by their constant dedication to solving the mysteries of dark matter and dark energy.

Global Collaboration and Discoveries

International collaboration in the field of space exploration is shown by Euclid. To make Euclid a reality, a group of member states, the European Space Agency (ESA), and the project as a whole have worked together. Each state has contributed its knowledge, resources, and finances to the mission.

International collaboration goes beyond the mission's planning and finance to include the scientific breakthroughs it promises. The data from Euclid have been analyzed and interpreted by scientists and researchers from all across the globe. Inspiring a worldwide community of astronomers and cosmologists committed to discovering the universe's most profound mysteries, this joint pursuit transcends geopolitical borders.

The broad-reaching effects of Euclid are evidence of the possibilities of worldwide cooperation in solving challenging astrophysical problems. The mission's findings have the potential to fundamentally alter our knowledge of the universe, and they are proof of the value of international scientific collaboration.

Anecdotes and Personal Narratives

The personal accounts and tales of the individuals who have devoted their lives and passions to the cause are hidden under the technical language and scientific terminology of Euclid. These tales give the project a human face, ranging from sleepless hours spent at observatories to exhilarating moments when data starts to flow.

The exhilaration of overcoming technical difficulties, the delight of seeing the first light collected by the telescope, and the camaraderie of working with coworkers from all backgrounds are just a few of the tales that scientists and engineers have told about their experiences working on Euclid.

Stories from the individual's life provide an insight into the commitment and tenacity needed to make space exploration a reality. The tremendous feeling of amazement and discovery that surrounds each achievement in the mission is also revealed by them.

Obstacles Overcome and Success

There have been difficulties along Euclid's trip. Technical, logistical, and operational challenges abound in space missions. From the difficult process of organizing worldwide partners to the demanding testing of equipment to verify their performance in the harsh environment of orbit, Euclid has its share of challenges.

The mission's delayed launch, which delayed the schedule for its scientific investigations, was one of the major difficulties it encountered. To overcome this delay, careful preparation, collaboration, and flexibility were necessary.

Euclid's mission victories have been as important. The successful launch of the spacecraft, the gathering of exact data, and the cooperation of global partners are enormous accomplishments. The science of astrophysics has already been impacted by the findings and insights produced by the mission.

The Euclid mission's emphasis on the human element of space exploration, in particular, underscores the commitment, teamwork, and individual stories that fuel scientific advancement. Euclid is a tribute to the strength of collective human effort in our quest to unravel the secrets of the cosmos as scientists, engineers, and multinational collaborators overcame obstacles and celebrated victories. To ensure that the human spirit of inquiry and discovery continues to light the universe as we look to the future, the legacy of Euclid acts as an inspiration for the subsequent generation of space explorers.

Chapter 9: Beyond Euclid: Exploring the Universe's Frontiers

Our knowledge of dark matter, dark energy, and the larger universe has already been significantly improved by the Euclid Space Telescope. The mission has, however, had an influence well beyond its original goals, ushering in a new age of astrophysical study and discovery. This chapter sets off on a voyage beyond Euclid, investigating the boundaries of the cosmos, unanswered questions, theoretical ramifications, and the far-reaching effects of Euclid's discoveries.

Unsolved Problems in Astrophysics

The cosmos continues to be filled with riddles and open questions, despite Euclid's great progress in illuminating the secrets of dark matter and dark energy. The observations of the mission have advanced our comprehension of these cosmic mysteries while also bringing to light fresh mysteries and oddities that contradict our preconceived notions.

The nature of dark matter particles, the possible evolution of dark energy throughout cosmic time, and the function of baryonic matter in the large-scale structure of the universe are just a few of the fascinating mysteries that yet remain in astrophysics. Euclid's statistics have shed a great deal of light on these issues, but solutions have eluded us so far.

The universe is also filled with mysteries that go beyond dark matter and dark energy. The nature of dark stars, the origins of cosmic magnetic fields, the presence of cosmic strings, and primordial black holes are just a few of the mysteries that astrophysicists hope to solve in the next few years.

Euclid's discoveries have theoretical ramifications.

The cosmology and astrophysics theoretical foundations have been significantly impacted by Euclid's discoveries. Existing hypotheses have been called into question by the data gathered by the telescope, which has also sparked the creation of fresh models to explain the discoveries.

Because of Euclid's exact observations of the rapid cosmic expansion, for instance, our knowledge of dark energy has undergone major refining. Discussions regarding the nature of dark energy itself have been prompted by these observations, which have consequences for theories of gravity such as adjustments to general relativity.

Similar to how the findings of Euclid have rekindled the study of dark matter. Even though dark matter is still a mystery, the mission's observations have helped to improve theories of its distribution and characteristics. Theoretical ramifications go beyond the Standard Model to include particle physics and maybe new physics as well.

Effects on Cosmology and Basic Physics

Beyond the narrower subject of astrophysics, Euclid's discoveries influence cosmology and basic physics. Cosmologists now have access to a wealth of information from the mission that is essential for comprehending the large-scale structure, history, and future of the universe.

Precision cosmology has benefited from Euclid's mission, which is one of its primary results. The data gathered by the telescope play a crucial role in limiting the cosmological parameters that characterize the universe's characteristics. Due to Euclid's discoveries, cosmic measurements have never been more precise, from the density of dark matter to the equation of state for dark energy.

Euclid's insights also have consequences for the foundations of physics. They test and improve our knowledge of space-time, gravity, and the structure of the cosmos. As we explore the furthest reaches of the universe, the interaction between cosmology and basic physics is expected to provide ground-breaking discoveries.

Imagining the Following Steps

The impact of the Euclid spacecraft extends beyond its original mission, acting as a launching pad for subsequent astronomical and space research projects. Future satellite telescopes and missions that will continue the search for the mysteries of the cosmos have been made possible by the data and findings from Euclid.

The James Webb Space Telescope (JWST), scheduled to launch soon, is one such project. By probing the infrared spectrum of the cosmos, JWST's sophisticated equipment and capabilities will add to Euclid's findings by revealing further details on dark matter, dark energy, and cosmic history.

Ongoing ground-based surveys and upcoming projects like the Square Kilometre Array (SKA) and the Large Synoptic Survey Telescope (LSST) are also expected to make substantial contributions to the study of astrophysics. To answer some of the most important questions about the universe, these initiatives will complement Euclid's objective.

The legacy of the Euclid expedition extends beyond the confines of a single mission, and its effects are felt across the whole discipline of astronomy. The universe's boundaries beckon as humans advance, driven by curiosity and equipped with cutting-edge technology, offering a future full of learning, comprehension, and wonder.

The astrophysics community has never been the same since the Euclid Space Telescope revealed the mysteries of dark matter and dark energy while sparking the flame of adventure. The questions that remain unanswered and the theoretical ramifications of Euclid's discoveries motivate us to push the frontiers of knowledge as we look into the shadow side of the cosmos. The effects on cosmology and basic physics are significant and will influence future space exploration. We picture a period where the universe exposes its mysteries one discovery at a time, using Euclid's legacy as our compass.

Conclusion

As the Euclid Space Telescope's mission comes to an end, we take a moment to consider the mission's profound significance, its contributions to our understanding of dark matter and dark energy, and the direction it has paved for future space exploration and cosmology.

Euclid Space Telescope Mission Reflections

The Euclid Space Telescope project is evidence of human inquisitiveness and inventiveness. It began an investigation into the secrets of dark matter and dark energy, two mysterious elements that determine the course of the cosmos. Through its voyage, Euclid has given us a wealth of information that has allowed us to see the universe more clearly and appreciate its beautiful pattern.

We now understand more about the big-picture structure of the universe thanks to the mission's exact observations of the accelerating cosmic expansion and the distribution of dark matter. It has shed light on the mysterious side of the cosmos, where these unobservable forces are in control and determine the course of galaxies and the whole universe.

A testimony to human accomplishment, Euclid's trip into space was the result of years of preparation, research, and teamwork. It has increased not just our knowledge of the cosmos but also global collaboration, proving the value of collaborative efforts in the search for knowledge.

The ongoing investigation into dark matter and dark energy

Dark matter and dark energy are still a cosmic mystery, even after Euclid made tremendous advances in our understanding of them. The results of the expedition have increased the intricacy of these phenomena and generated additional concerns for researchers to consider.

More research is being done on the characteristics and interactions of dark matter since its nature continues to defy us. Our knowledge of gravity and the underlying structure of the cosmos is put to the test by the behavior of dark energy as revealed by Euclid's data. The universe is brimming with mysteries that beg us to delve deeper and test the limits of our understanding.

The continuous story of discovering the mysteries of the cosmos includes Euclid's endeavor. It is evidence of our never-ending need for knowledge and our everlasting dedication to comprehending the universe that surrounds us. Our curiosity will be piqued and our desire to learn more will be fueled by the mysteries of dark matter and dark energy.

Cosmology and Space Telescopes in the Future

The Euclid Space Telescope's impact goes beyond its original purpose. It motivates cosmology and space exploration in the future. The expedition has shown how cutting-edge technology and worldwide cooperation may increase our knowledge of the cosmos.

Future missions like the James Webb Space Telescope (JWST) and others are prepared to build on Euclid's accomplishments. JWST's infrared capabilities will supplement Euclid's observations and enable further exploration of the universe's mysteries. The boundaries of astrophysics will keep growing thanks to current ground-based surveys and the next research.

The cosmos urges us to continue exploring the unknown with its wide stretches and well-kept mysteries. Future space telescopes and cosmological projects may follow in the footsteps of Euclid's expedition, which has served as a guiding beacon. We are eager to explore the secrets that lie ahead as we gaze upwards, curious and filled with a deep feeling of wonder.

The Euclid Space Telescope project has permanently altered our perception of the universe. The limits of knowledge have been stretched, the mysteries of dark matter and dark energy have been revealed, and a love for exploration has been sparked. We are on the verge of a new period in human history of discovery as we reflect on the unsolved riddles and look to the future of space telescopes. The Euclid Space Telescope has set the stage, and the universe's big show is still going strong while waiting for human involvement in the continuing effort to solve some of its most profound riddles.

Printed in Great Britain
by Amazon